Mixtures and Solutions

Full Option Science System
Developed at
The Lawrence Hall of Science,
University of California, Berkeley
Published and distributed by
Delta Education,
a member of the School Specialty Family

1487711
978-1-62571-360-5
Printing 2 – 6/2016
Webcrafters, Madison, WI

Table of Contents

Mixtures

Sand is a mixture.

The next time you are at the beach, pick up a handful of sand. Look at it closely. You will see that each grain of sand is really a tiny rock. Not only that, but the grains are different colors. And some of the grains are not rocks at all. They are little pieces of seashells.

Sand is a **mixture**. A mixture is two or more materials put together, or combined. This sand is a mixture of black, white, tan, and gray rocks, and bits of shell. A handful of sand is a mixture of several different things.

Mixtures are everywhere. But if you are not looking for them, you could miss them. The sidewalk is a mixture you can walk on. It is several sizes of rock mixed with cement. A bowl of mixed nuts is a mixture you can eat. So are fruit salad, vegetable soup, and carrot-and-raisin salad. And if you ate some mixed nuts, vegetable soup, and salad, just think about the mixture in your stomach!

A mixture of nuts

A mixture of vegetables and water (soup)

A mixture of carrots and raisins (salad)

A ball stays the same size and shape in any container.

Making Mixtures

Mixtures like sand or mixed nuts are made when two or more solid materials are combined. **Solid** is one of the three common states of **matter**. Solid objects have **mass**, take up space, and have a definite shape and **volume**. A ball, for instance, stays the same size and shape in your hand, on a tabletop, or in a cup.

Chocolate syrup poured into milk is a mixture. Lemon juice and water is a mixture. So is oil and vinegar. These are examples of mixtures made of two **liquid** materials. Liquid is the second common state of matter. Liquids have mass, take up space, and have a definite volume. But liquids do not have a definite shape. A volume of water can have a different shape depending on whether it is in your hand, on a tabletop, or in a cup. Liquids take the shape of the container they are in.

Your breath is a mixture. The exhaust coming out of a car is a mixture. The air that surrounds Earth is a mixture. These are examples of mixtures made of **gaseous** materials. **Gas** is the third common state of matter, but we can't always see gases. Most gases are colorless and **transparent**. Some gases, however, have color, like those that make smog. We can see air on a smoggy day.

Water changes shape depending on the container you put it in.

A smoggy
day in
New York City

Gases have mass and take up space, but they do not have definite volume or shape. A mass of air will not stay in your hand, on a tabletop, or in a cup. Gases are shapeless and expand to fill any closed container they are in.

Solids and liquids are often mixed. **Salt** and pepper are mixed with oil and vinegar to add flavor to salad dressing. Flour and water are mixed to make bread. Cereal and milk are mixed for breakfast. Rice is mixed with water to cook it.

Gases and liquids are mixed sometimes, but often the gas separates from the liquid as bubbles. That's what happens when you pour root beer into a glass. The **carbon dioxide** (CO_2) gas that was mixed with the liquid root beer forms bubbles. The bubbles rise to the surface and pop. Then the carbon dioxide in the bubbles mixes with the air. But for a while, the root beer is a lively mixture of liquid and gas.

What ingredients make up this breakfast mixture?

What is the liquid and what is the gas in this mixture?

5

Mixtures of solid and gas aren't often made on purpose. But they happen all the time by accident. If you fill a glass with marbles, you have a mixture of marbles and air. The spaces between the marbles are filled with air. The same is true of any solid object in a container. When it snows, the air (gas) is mixed with frozen water (solid). Dust floating around in the air is also a mixture of solid and gas.

Can you have a mixture of solid, liquid, and gas? Yes. Remember that glass of root beer? Just add a few ice cubes. The glass of ice-cold root beer is a mixture of solid, liquid, and gas.

Mixing Solids and Liquids

Mixtures of solids and liquids are interesting. Several things can happen. When gravel and water are mixed, the gravel sinks to the bottom of the container. If you stir the mixture, things move around, but that's about it.

Mixing gravel and water

Mixing powdered milk and water

Mixing salt and water

Gravel mixture after 5 minutes

Powdered milk mixture after 5 minutes

Salt mixture after 5 minutes

When you mix powdered milk and water, the powder makes the mixture cloudy white. The mixture remains cloudy.

When you mix salt and water, the salt disappears. The mixture is transparent and colorless.

Gravel, powdered milk, and salt all make mixtures with water. After stirring, you can still see the gravel and milk, but the salt is gone. Salt is different in some way.

A mixture of salt and water forms a **solution**. When solid salt and liquid water are mixed, the solid disappears into the liquid, and the mixture is transparent. This mixture is a solution. A solution is a special kind of mixture.

When a solid material disappears in a liquid, it is not gone. It has **dissolved**. When the solid material dissolves, it breaks into pieces so tiny that they are invisible. The solid material that dissolves is called the **solute**. The liquid in which the solute dissolves is called the **solvent**. Every solution is made of a solute dissolved in a solvent. Salt (solute) dissolves in water (solvent) to make a saltwater solution.

Salt (solute) dissolves in water (solvent).

The Universal Solvent

Water dissolves thousands of **substances**. For this reason, it is called the universal solvent. Countless substances are dissolved in the oceans. The fluids in living organisms, such as blood and sap, contain water with thousands of substances dissolved in the water. Look around. Does everything dissolve in water?

Water Properties

Boiling point = 100° Celsius (C)

Freezing point = 0° Celsius (C)

Density = 1 gram per milliliter (g/mL)

Taking Mixtures Apart

Y ou made a mixture of gravel and water in class. You put 50 milliliters (mL) of water and a spoonful of gravel in a cup, and the job was done. Then you separated the mixture of gravel and water. You poured the mixture through a screen. The gravel stayed on the screen, and the water passed through.

All mixtures can be separated. But not all mixtures can be separated in the same way. The **physical properties** of the materials in the mixture can be used to separate the mixture.

Particle size is a physical property of gravel. Particle size is a physical property of water. The particles of gravel are larger than the holes in the screen. The particles of water are smaller than the holes in the screen. The screen can be used to separate the mixture.

In class, the mixture of **diatomaceous earth** and water passed through the screen. The particles of the powder and water are both smaller than the holes in the screen. What property will separate powder from water? The answer is size again. Powder particles are larger than the holes in filter paper. Water particles are smaller. Filter paper will separate a mixture of powder and water.

A screen can separate gravel and water.

Other Ways to Separate Mixtures

Imagine you open a drawer to get a rubber band. Oops, the rubber bands spill. So do a box of toothpicks and some nails. The drawer has an accidental mixture of nails, toothpicks, and rubber bands. How can you separate the mixture?

You can use the property of shape. You can pick out each piece one at a time. But it might take 10 minutes to separate the mixture.

Nails are made of steel. Steel has a useful property. Steel sticks to magnets. If you have a magnet, you can separate the steel nails from the mixture in a few seconds. **Magnetism** is a property that can help separate mixtures.

A mixture of nails, toothpicks, and rubber bands

What about the toothpicks and rubber bands? Wood floats in water. Rubber sinks in water. The property of **density** (sinking and floating) can be used to separate the wood toothpicks and rubber bands. Drop the mixture into a cup of water. Then scoop up the toothpicks from the surface of the water. Pour the water and rubber bands through a screen. The water will pass through the screen, but the rubber bands won't. Now the job is done.

Magnetism is a property that helps separate mixtures.

Density is a property that helps separate mixtures.

9

Separating Solutions

A mixture of salt and water is a solution. The dissolved salt particles and the water particles are both smaller than the holes in filter paper. The property of size is not useful for separating a solution of salt and water. But **evaporation** will work.

Evaporation is the **change** of state from liquid to gas.

Salt crystals

Water **evaporates**, but salt does not. When a salt solution is left in an open container, the water slowly turns to gas and goes into the air. The salt is left behind. Solutions can be separated by evaporating the liquid.

The salt left behind after evaporation doesn't look like the salt that dissolved in the water. Is it still salt? Yes, it is. When the water evaporates, the salt reappears as salt **crystals**. Salt crystals always look square. Salt crystals often have lines going from corner to corner, forming an X.

Many solid materials dissolve in water to make solutions. When the water evaporates, the materials reappear as crystals. Each different material has its own crystal shape. Some crystals are needle shaped. Other crystals are six sided. Others are like tiny fans.

Crystal shape is a physical property. Crystal shape can be used to identify materials. Whenever you observe square shaped crystals in an evaporation dish, you will know that salt might be one of the ingredients in the solution.

Three different crystal forms

Conservation of Matter

There's one more thing to think about when you make a mixture. All matter has mass. Anything that has mass is matter. If you have 50 milliliters (mL) of water in a cup and add 30 grams (g) of sand to the cup of water, what will the mass of the mixture be? The mass of the 50 mL of water is 50 g, so the mass of the mixture will be 50 g (water) + 30 g (sand) = 80 g (mixture). That seems pretty easy to understand.

50 g water   **+**   **30 g sand**   **=**   **80 g mixture**

But the mixture of salt and water is a little trickier to think about. When you mix 30 g of salt with 50 g of water, what do you think the mass of the mixture will be? The salt disappears in the water, so what happens to its mass? Did you conduct this investigation? The mass of the clear solution is 80 g. That is the sum of the mass of the water (50 g) and the mass of the salt (30 g). The mass of a substance like salt does not change when it dissolves. Even if you can't see it, the salt is still there, and its mass has not changed. In fact, mass never goes away. Mass is **conserved**, therefore matter is conserved. That means matter can change shape, state, or location, but it can never be lost or destroyed.

50 g water   **+**   **30 g salt**   **=**   **80 g salt mixture**

Matter is never destroyed, but it can change. Wood (matter) changes to ash when it burns.

Sometimes it is hard to understand how matter is conserved. For instance, when you have a campfire, a large mass of wood burns and all that is left at the end of the evening is a small pile of ash.

If matter is conserved, where did the mass of the wood go? The fire produced several things. It produced smoke, light, and heat. Light and heat are **energy**. Energy is not matter. Smoke is gas and tiny particles of soot. Gas and soot are matter. That's where the wood went. The fire changed most of the mass of the wood into gas and tiny particles. The particles drifted off into the air. Gases and tiny particles have mass.

If you could capture all the smoke and dust coming up from the fire, and gather up all the ashes, what would you find? You would find that the mass of the gas and ash would add up to the mass of the wood you put on the fire earlier. Conservation of matter is just one of the great truths of nature. Matter can never be destroyed, but it can be changed.

As you continue your investigations of mixtures, you might find an instance where your observations suggest that matter is not conserved. But matter is always conserved. You will have to do some deep thinking to explain why your evidence suggests that matter is conserved.

The Story of Salt

Most people don't notice it, but salt is everywhere. It is in most of the foods we eat. It is in ocean water and used in cooking. There are places on land where great deposits of salt are found. The salt flats near Salt Lake City, Utah, are one example. That salt was left behind when seawater evaporated a very long time ago.

Salt (**sodium chloride**) has been important to people since early times. It was often used to keep food from spoiling. Heavily salted food could be preserved for a long time. When sailors spent months at sea, the foods they carried on their ships were heavily salted.

Salt is the oldest-known food additive. The earliest reference to salt was written in China around 2700 BCE. It described how salt made food taste better and helped people stay healthy.

The value of salt led to the development of the salt industry. Salt was first mined from salt deposits on land. It was packaged and sold or traded in marketplaces. In Europe, some monarchs placed heavy taxes on salt. In France, the salt tax was one of the reasons for the French Revolution in 1789.

Salt was important in United States history, too. Salt became one of the main products shipped along the Erie Canal in the 1800s. During the Civil War (1861–1865), Union forces attempted to destroy the South's salt industry. They did not want the South to be able to preserve food for its soldiers.

Today, more uses have been found for salt. It is added to the diets of livestock to improve their health. It is used to soften hard water and to **melt** ice on slippery roads. Salt is also used to make other chemicals, such as chlorine and sodium.

Science Practices

1. **Asking questions.** Scientists ask questions to guide their investigations. This helps them learn more about how the world works.

2. **Developing and using models.** Scientists develop models to represent how things work and to test their explanations.

3. **Planning and carrying out investigations.** Scientists plan and conduct investigations in the field and in laboratories. Their goal is to collect data that test their explanations.

4. **Analyzing and interpreting data.** Patterns and trends in data are not always obvious. Scientists make tables and graphs. They use statistical analysis to look for patterns.

5. **Using mathematics and computational thinking.** Scientists measure physical properties. They use computation and math to analyze data. They use mathematics to construct simulations, solve equations, and represent different variables.

6. **Constructing explanations.** Scientists construct explanations based on observations and data. An explanation becomes an accepted theory when there are many pieces of evidence to support it.

7. **Engaging in argument from evidence.** Scientists use argumentation to listen to, compare, and evaluate all possible explanations. Then they decide which best explains natural phenomena.

8. **Obtaining, evaluating, and communicating information.** Scientists must be able to communicate clearly. They must evaluate others' ideas. They must convince others to agree with their theories.

Scientists ask questions and communicate information. Are you a scientist?

Engineering Practices

1. **Defining problems.** Engineers ask questions to make sure they understand problems they are trying to solve. They need to understand the constraints that are placed on their designs.

2. **Developing and using models.** Engineers develop and use models to represent systems they are designing. Then they test their models before building the actual object or structure.

3. **Planning and carrying out investigations.** Engineers plan and conduct investigations. They need to make sure that their designed systems are durable, effective, and efficient.

4. **Analyzing and interpreting data.** Engineers collect and analyze data when they test their designs. They compare different solutions. They use the data to make sure that they match the given criteria and constraints.

5. **Using mathematics and computational thinking.** Engineers measure physical properties. They use computation and math to analyze data. They use mathematics to construct simulations, solve equations, and represent different variables.

6. **Designing solutions.** Engineers find solutions. They propose solutions based on desired function, cost, safety, how good it looks, and meeting legal requirements.

7. **Engaging in argument from evidence.** Engineers use argumentation to listen to, compare, and evaluate all possible ideas and methods to solve a problem.

8. **Obtaining, evaluating, and communicating information.** Engineers must be able to communicate clearly. They must evaluate other's ideas. They must convince others of the merits of their designs.

Engineers use models.

Extracts

One of the most popular beverages in the world is tea. Tea is a plant grown all over Asia. The leaves are picked by hand, dried, and shipped around the world. Tea is prepared for drinking by soaking tea leaves in boiling water.

Another beverage enjoyed all over the world is coffee. Coffee is the seed of the coffee tree fruit. The seeds, called beans, are removed from the fruit, dried, and then roasted. The roasted beans are ground into a coarse powder. Boiling water is poured over the ground beans to prepare the coffee.

Tea and coffee are both **extracts**. An extract is a solution. To make an extract, plant material (leaves, bark, roots, seeds, flowers, and so on) is put into a solvent. The solvent dissolves *some* of the substances out of the plant material. The plant material doesn't dissolve completely. Only a tiny part of it dissolves. Often the only evidence that anything has dissolved is a change to a solvent. Extracts contain dissolved substances that have color, odor, and taste.

You probably know of some other extracts. Vanilla is an extract made by soaking vanilla beans (a seedpod) in a solvent. The flavor of root beer is an extract made by soaking the bark from the roots of sassafras trees in a solvent. Peppermint, wintergreen, almond, and many other flavors are used to prepare foods and drinks. Extracts allow us to enjoy the special flavors of these plants without having to work with the plants that produce them.

Many of our favorite tastes are oils. Cinnamon, peppermint, vanilla, and many others are oils in the host plants. We know about oil and water. They don't mix. This means the peppermint taste won't dissolve in water. How do we get the flavor in an extract? We use another solvent. Oils do dissolve in ethanol, so ethanol is used to extract many of our favorite flavors. The ethanol can then be dissolved in water to use the extract in cooking.

Some extracts are made using oil as the solvent. People who like really spicy tastes can get red hot pepper oils. Ground hot peppers are put into light oil, like sesame oil. The oil is a solvent for the really hot substances in the peppers.

The perfume industry relies on extracts. Extracts of roses, orange blossoms, honeysuckle, and lilac are a few of the wonderful scents used for perfume.

Beachcombing Science

An accident in the Pacific Ocean led to an unusual opportunity for **oceanographers**. Oceanographers are scientists who study ocean and wind currents. A storm struck a cargo ship on its way from Hong Kong to Tacoma, Washington, on January 10, 1992. High seas washed a shipping container carrying 29,000 bathtub toys overboard south of the Aleutian Islands. Soon plastic turtles, frogs, beavers, and ducks were floating in the ocean. Ten months after the accident, the toys began to wash up on beaches near Sitka, Alaska.

Scientists at the National Oceanic and Atmospheric Administration (NOAA) took advantage of this opportunity. A group of oceanographers, including Curtis Ebbesmeyer and Jim Ingraham, began to track the toys' journeys around the Pacific Ocean. At first, they followed reports from beachcombers who found many of the toys washed up on beaches. Then they actively went out and searched for the toys themselves. They made charts showing where, when, and how many toys washed up. These observations gave them direct evidence about ocean and wind currents. Scientists hoped to use this information to update existing **models** of ocean currents. With accurate models of currents, scientists would be better prepared to predict the movement of things in the ocean, such as lost sailors or oil spills.

The toys fell off a ship in the North Pacific in 1992 and made their way to southeast Alaska, near Sitka.

The oceanographers learned some surprising things. According to Ingraham, scientists thought it would take from 4 to 6 years for the toys to return to where the shipping container fell into the ocean. Instead, it took just 2.5 years. That might mean that ocean and wind currents are stronger than scientists thought. Scientists also noted that the toys traveled in a counterclockwise path.

Even though the toys washed up on beaches in the northern Pacific, many of them are still out there. Ingraham wouldn't be surprised if some of the toys become trapped in northward-moving ice at the North Pole. If that happened, then those toys might make it all the way to the Atlantic Ocean. That would give scientists an opportunity to gather data about currents in another part of the ocean, all thanks to a container of shipwrecked toys!

The current that carried the bathtub toys around the North Pacific is the Subpolar Gyre (rhymes with *fire*). The bathtub-toy observations improved a computer model of ocean surface currents developed by Jim Ingraham. The Subpolar Gyre model was first developed with data from a Messages in Bottles (MIB) project. The MIB project released 33,869 bottles into the ocean in the 1950s. The beachcombing data collected from the turtles, ducks, beavers, and frogs that washed up near Sitka, Alaska, confirmed the model suggested by the MIB project.

Other data have described a second, larger Pacific gyre, the Subtropical Gyre. The Subtropical Gyre has trapped a Texas-size garbage patch of tiny bits of plastic from decomposing trash.

Map of the North Pacific showing two rotating currents, or gyres

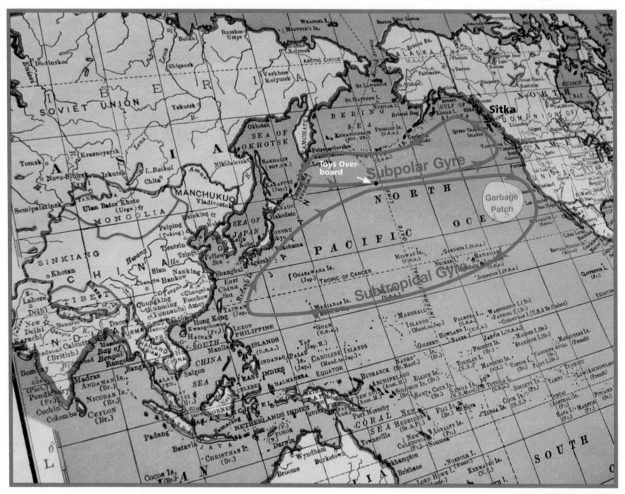

Solid to Liquid

Ice melts. It changes from solid to liquid. An ice cube in a cup on your desk will change into water in about an hour. Chocolate and butter melt, too. But they will not melt on your desk. You have to put them in hot water to make them melt. Wax melts a little bit in hot water. It gets soft. But a pebble won't melt at all. Or will it?

What causes the butter to melt? Heat energy. If you put butter in a cup, nothing happens right away. If you put the cup in hot water, the butter melts. Heat energy **transfers** from the hot water to the butter. Heat energy makes the butter melt.

Heat energy is causing this butter to melt.

But ice melts without heat energy. Why is that? Actually, heat energy does make ice melt. When ice is in the freezer, it doesn't melt. It stays solid, or frozen. When you bring ice out into a room that is warmer than the freezer, the ice melts. That's because heat energy from the room transfers to the ice and causes it to melt.

Solid ice melts to form liquid water.

Materials melt at different **temperatures**. Water melts at 0°C. Water **freezes** at 0°C, too. When water is below 0°C, it is solid. When it is above 0°C, it is liquid. Chocolate melts at about 50°C. Candle wax melts into liquid at around 80°C. And yes, the pebble will melt when it is heated to over 1,000°C! Have you ever seen lava flowing from a volcano? That's melted rock.

Metals melt, too. Jewelers melt gold and silver to make rings and other beautiful things. Sculptors melt bronze to make statues. Iron and copper are melted to separate them from the ores taken from mines. Sand is melted to make glass. Many things that we think are always solid will melt if enough heat energy is transferred to them.

Lava flowing down the side of Kilauea Volcano in Hawaii

Gold melts at 1,064°C.

Liquid and Gas Changes

When something is wet, it is covered with water, or it has soaked up a lot of water. When it **rains**, everything outside gets wet. When you go swimming, you and your swimsuit get wet. Clothes are wet when they come out of the washer, and a dog is wet after a bath.

But things don't stay wet forever. Things get dry, often by themselves. An hour or two after the rain stops, stairs, sidewalks, and plants are dry. After a break from swimming to eat lunch, you and your swimsuit are dry. After a few hours on the clothesline, clothes are dry. A dog is dry and fluffy a short time after its bath. Where does the water go?

It evaporates. When water evaporates, it changes from liquid water to **water vapor**, a gas. The gas drifts away in the air.

Wet stairs just after a rainstorm   **Dry stairs the next day**

What causes liquids to evaporate? Heat. When enough heat transfers to a liquid, the liquid changes into a gas. The water vapor leaves the wet object and goes into the air. As the water evaporates, the wet object gets dry.

What happens when you put a wet object in a sealed container? It stays wet. If you put a wet towel in a plastic bag, it's still wet when you take it out of the bag. Why? A little bit of the water in the towel evaporates, but it can't escape into the air. The water vapor has no place to go, so the towel stays wet.

Have you ever seen water vapor in the air? No, water vapor is invisible. When water changes into vapor, it changes into individual water particles. Water particles are too small to be seen with your eyes. The water particles move into the air among the **nitrogen** and **oxygen** particles. Water vapor becomes part of the air. When water becomes part of the air, it is no longer liquid water.

Gas to Liquid

What happens to all that water in the air? As long as the air stays warm, the water stays in the air as water vapor. But when air cools, the water particles start to come together. Tiny droplets of liquid water form.

When gas particles come together to form liquid, it is called **condensation**. Condensation is the change from gas to liquid. When water condenses, it becomes visible again. We see the condensed water as clouds, fog, and dew.

What Else Evaporates?

Water isn't the only liquid that evaporates. Gasoline for cars is a liquid. It is kept in tightly closed gas tanks. If the gasoline was left in open containers, it would evaporate and disappear into the air.

Here's an interesting fact. We breathe oxygen from the air. Oxygen in the air is a gas. But if you put a container of oxygen in a freezer that is –183°C, the oxygen becomes liquid. If there was any liquid oxygen on Earth, it would evaporate because Earth is much warmer than –183°C.

Water vapor condenses indoors, too. On a cold morning, you might see condensation on your kitchen window. Or if you go outside into the cold wearing your glasses, they will get fogged with condensation when you go back inside.

What happens to the bathroom mirror after you take a shower? The air in the bathroom is warm and filled with water vapor. When the air touches the cool mirror, the water vapor condenses on the smooth surface. That's why the mirror gets foggy and wet.

Condensation on a window

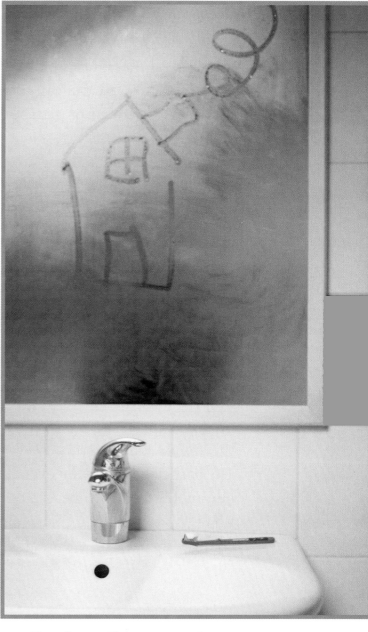

Condensation on a mirror

Solutions Up Close

Salt solutions are transparent. You can't see anything in them. When you look at a salt solution with a hand lens, what do you see? Still nothing. In fact, you can't see anything in a salt solution even with the most powerful light microscope. Does that mean the salt is gone when it dissolves?

No, the salt is still there. To understand what happens to the salt, you have to think very small. You have to think about pieces of salt so small that it takes billions and billions of them to make a tiny salt crystal. We can call the tiniest piece of salt a salt particle.

Water is also made of particles. Water particles are different from salt particles, but they are about the same size. In liquid water, the particles are always moving around and over one another.

Let's imagine that we can see the salt particles. We'll represent one salt particle with this pink circle.

One tiny crystal of salt might look like this. The salt crystal is a solid. It has mass, shape, and a definite volume.

Here is a container of water a million times smaller than a 5-milliliter (mL) spoonful. If we could see water particles, they might look like this. Water is a liquid. The particles are moving over and around one another all the time. That's how water flows. Water has a definite volume, but it changes shape to fit its container.

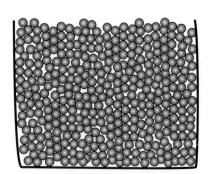

When you put a crystal of salt in a container of water, the salt sinks to the bottom.

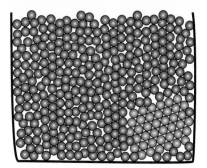

The particles of water bump into the salt crystal. This action knocks salt particles off the crystal. The loose salt particles become surrounded by water particles.

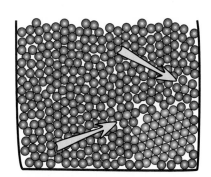

The salt particles are carried into the water. They end up spread evenly among the water particles. The particles of salt among the water particles are the dissolved salt. The particles of salt still on the bottom of the container are undissolved salt crystals.

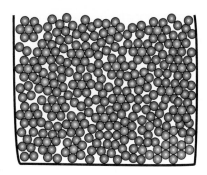

A solution forms when salt particles break free from the salt crystal and get carried away by the water particles. The breaking away and getting carried away is the process of dissolving.

Interpreting the Diagrams

Work with a partner. Look at the diagrams on pages 26–27. Describe to your partner what happens when salt dissolves in water.

Concentrated Solutions

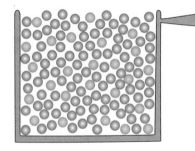

If you stop by the freezer section at the market, you can pick up a can of orange juice. But be careful! If you defrost it and try to drink the orange juice straight from the can, you will be in for a shock. The orange juice is thick and strong. That's because the can contains orange juice concentrate. Most of the water has been removed. The juice is too **concentrated** to drink.

Orange juice straight out of the orange is a solution. Water is the solvent. At the orange juice factory, the orange juice is heated. Water particles leave the solution, but orange juice particles don't. The orange juice particles are still evenly distributed in the water, but there is less water. As a result, the orange juice particles get closer together, because there are fewer water particles between them. Here's how that works.

1. Here is a pot of orange juice. There are 100 water particles (blue) and 25 orange juice particles (orange).

   A representation of orange juice in a pot

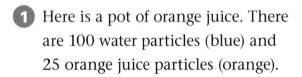

2. The pot of juice is slowly heated. Water particles begin to evaporate.

   Water particles evaporate when juice is heated.

3 When 75 water particles have evaporated, the orange juice looks like this.

The pot of orange juice after most of the water has evaporated

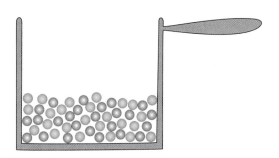

4 Look at the pots before and after evaporation. What is the same? What is different?

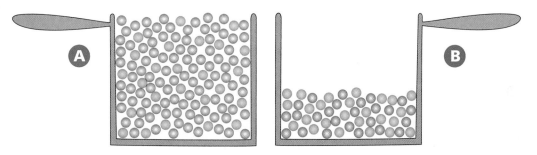

A comparison of a pot of orange juice before (A) and after (B) heating

Both pots contain solutions made with the same materials: water particles and orange juice particles. Both pots also have the same number of orange juice particles.

The important difference is the amount of water. The fresh juice in Pot A has 100 water particles. The evaporated juice in Pot B has only 25 water particles. The orange juice in Pot B is *more concentrated*.

That means that if you scoop up 100 milliliters (mL) of concentrated solution from Pot B, it will contain more orange juice particles than 100 mL of solution from Pot A.

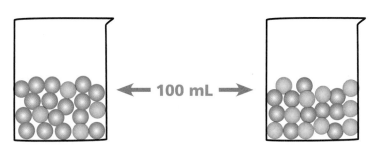

Orange juice from Pot A   **Orange juice from Pot B**

You can also think of **concentration** as the **ratio** of water particles to orange juice particles. There are four times as many water particles as orange juice particles in Pot A. The ratio of water to orange juice is four to one. In math that is written 4:1.

In Pot B there are 25 water particles and 25 orange juice particles. The ratio of water to orange juice is 1:1. There is only one water particle for each orange juice particle in Pot B. So that solution is *more concentrated* than the solution in Pot A. Pot A has four water particles for each orange juice particle.

Comparing Concentrations

If you have two different solutions made with the same materials, such as salt and water, you can often use a balance to figure out which one is more concentrated. Here's how.

- Measure equal volumes of Solution X and Solution Y.
- Weigh them.
- The sample with the greater mass is more concentrated. It's just that simple!

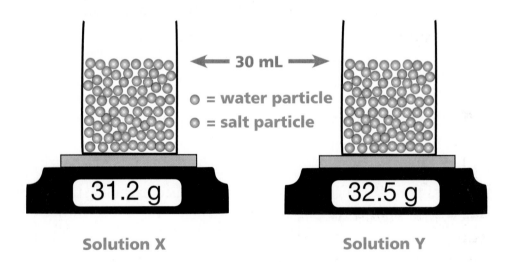

Solution X   Solution Y

How does that work? Particles have mass. Salt particles have mass, and water particles have mass. But here's the important thing: each salt particle has more mass than each water particle. When salt particles go into solution, they take up space. They push some water particles out of the way. When you take a volume of a concentrated salt solution, there are more heavy salt particles in it than there are in an equal volume of a **dilute** salt solution. So when you compare equal volumes of the two salt solutions, one will have more mass because it has more salt particles. And the one with more salt particles is more concentrated.

Thinking about Concentration

1. Which solution on page 30 is more concentrated, Solution X or Solution Y?

2. Look at the four salt solutions below (A, B, C, and D). Put them in order from most concentrated to most dilute.

Each sample on the balance is exactly 150 mL.

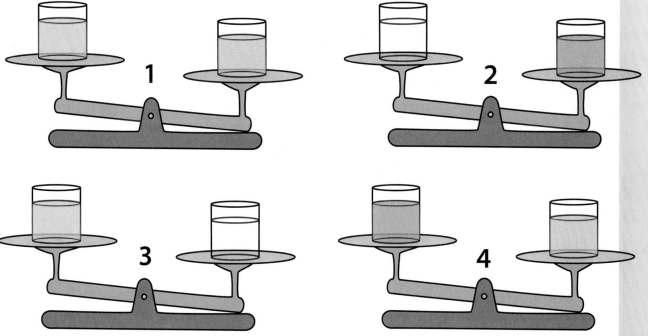

The Air

Air is a mixture of gases. The most common gas in air is nitrogen. Air is about 78 percent nitrogen. The second most common gas is oxygen. Air is about 21 percent oxygen. The other 1 percent is composed of dozens of other gases, with a bunch of tiny solids floating around. Air is a complex mixture.

Another way to look at air is as a solution. Think of air as a bunch of gas particles dissolved in nitrogen. The most concentrated gas dissolved in the nitrogen is oxygen. All other gases are part of the solution, but they are present in low concentrations.

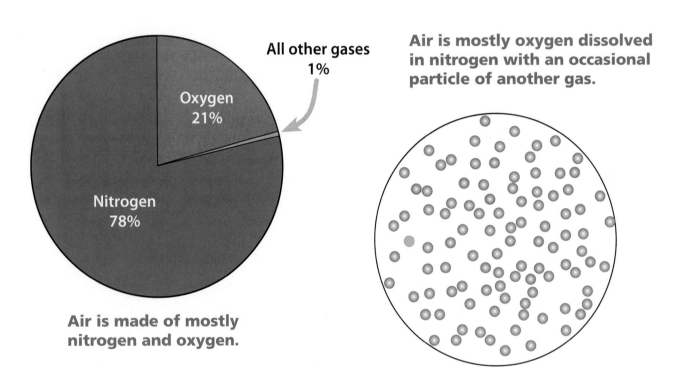

All other gases 1%

Oxygen 21%

Nitrogen 78%

Air is made of mostly nitrogen and oxygen.

Air is mostly oxygen dissolved in nitrogen with an occasional particle of another gas.

Two Other Gases in the Air

You might know the names of a couple of other important gases dissolved in the nitrogen: water vapor and carbon dioxide (CO_2).

When water evaporates, where does it go? It becomes part of the air. Individual water particles change into gas and dissolve in the air. On a warm, humid day, the concentration of water vapor in the air might reach 4 percent. Even so, we never see it. Water in the gas state is invisible.

We don't see water vapor until it cools and condenses. Condensed water vapor is liquid. The tiny liquid droplets form visible clouds. We recognize these processes of water entering the air (evaporation) and leaving the air (condensation) as part of the water cycle.

Water vapor condenses in the air to form clouds.

Fuel is a source of energy. When fuels, like wood, coal, grass, and peanut butter sandwiches, are burned, waste gases are released into the air. One of the most concentrated of them is CO_2. The concentration is low (0.04 percent), but it is significant. Carbon dioxide comes from forest-fire smoke, car exhaust, and the breath you exhale. Carbon dioxide in the air absorbs energy from sunlight. Scientists are concerned about the concentration of CO_2 in the air. Rising CO_2 concentration leads to rising temperatures worldwide.

Air in Other Places

The International Space Station orbits Earth high in the **atmosphere**, about 340 kilometers (km) above Earth's surface. The "air" up there is about 28 percent oxygen and 71 percent helium. The nitrogen is less than 1 percent. The concentration of gases is extremely low. How do astronauts survive in this environment? What do they breathe?

The air inside the space station is just about the same as it is on Earth's surface. The astronauts breathe a mixture of nitrogen and oxygen brought into space in pressurized bottles. Astronauts don't wear space suits inside. But when they go outside, they do wear a protective space suit. The suit is filled with pure oxygen—no nitrogen.

Living and working in small spaces creates a problem. Astronauts exhale CO_2. Carbon dioxide is toxic. A concentration of 1 percent CO_2 makes a person sleepy. A concentration of 7 percent causes dizziness, headache, and blackout. In order to keep the air safe for the astronauts, special filters are used to remove CO_2 from the air supply.

Oceans cover 71 percent of Earth's surface. Scientists explore the deep ocean water in submarines. The space inside is supplied with a mixture of gases similar to air on Earth's surface. But in shallower water, up to 100 meters (m), people dive in and swim around with a supply of air to breathe. Plain pressurized air is fine if you don't go deeper than 30 m. If you go deeper, **pressure compresses** the gas, increasing the concentration of the oxygen and nitrogen. Below 30 m, the pressure makes nitrogen toxic, which causes people to make bad decisions.

To remedy the problem, divers mix oxygen with helium. Helium doesn't affect thinking, so divers can go deeper. A mixture of 80 percent helium and 20 percent oxygen allows divers to go below 30 m safely for short periods of time.

Famous Scientists

Joseph Priestley

Joseph Priestley (1733–1804) was born in England. He grew up expecting to become a minister. But that was only one of his many interests. He was a teacher, a philosopher, and a scientist.

Priestley conducted experiments with electricity. He was the first to describe how electric force worked. This work was the basis for important work of many other scientists.

In the 1770s, Priestley became interested in gases, which he called airs. His experiments led to the discovery of sulfur dioxide, nitric oxide, nitrous oxide, ammonia, and perhaps oxygen. The "perhaps" is because it isn't clear who first isolated oxygen.

Priestley invented soda water. He used pressurized carbon dioxide to saturate water. But he failed to see the commercial value of his invention. J. J. Schweppe took Priestley's idea and made a fortune.

Antoine Lavoisier

Antoine Lavoisier (1743–1794) was born in France. He was well educated and became a lawyer. His second calling was science.

Lavoisier burned things. Most scientists thought things burned because they contained phlogiston, a firelike substance. Lavoisier doubted the existence of phlogiston. Here's how he set out to disprove it.

Most people thought that burning released phlogiston and that's why the ash had less mass. Lavoisier "burned" mercury in a closed container with air. After a couple of hours, the mercury shrank to reddish "ash." The volume of air was smaller, too. Further, the mass of the "ash" was greater than the mass of the starting mercury.

Lavoisier figured out that oxygen in the air reacted with the mercury. This formed mercury oxide. The mercury oxide was heavier than the mercury. Lavoisier went on to describe a new way to think about **chemical reactions**. The phlogiston theory started to lose favor. Because of this work, Lavoisier is sometimes called the father of modern chemistry.

Marie Curie

Maria Sklodowska (1867–1934) was born in Warsaw, Poland. Her father was a science and math teacher. He taught his children basic laboratory science methods. When she was 10 years old, Curie went to boarding school. She graduated at 16 with a gold medal. She was unable to go to college because she was a girl. For a while, she was a governess, teaching young children.

In 1890, Curie moved to Paris. There, she became known as Marie. She studied math, physics, and chemistry on her own. In 1891, she enrolled in the University of Paris. After graduating, she investigated the magnetic properties of steel. She did experiments in the laboratory of Pierre Curie. Pierre convinced Curie to continue studying at the university. Soon after that, Pierre and Curie married.

In 1895 and 1896, physicists discovered powerful radiation called X-rays. The element uranium produces radiation like X-rays. Curie became interested in uranium radiation. She called it **radioactivity**. She investigated many sources of radioactivity. She found several minerals with much stronger radioactivity than uranium. Finally, she isolated the sources of the strong radioactivity. They were two unknown elements. Polonium was named for Curie's home country. The other element was named radium.

Curie's life was very productive. She eventually received two Nobel Prizes. In 1903, she received the Nobel Prize in Physics. That prize honored her pioneering work on radioactivity. In 1911, she received the Nobel Prize in Chemistry. That prize was for discovering polonium and radium. In 1934, Marie Curie died from a blood disease. Her long-term exposure to radiation may have caused her death.

Carbon Dioxide Concentration in the Air

When Charles David Keeling (1928–2005) started working at Caltech University in 1953, he was interested in how carbon dioxide (CO_2) moved around in the environment. How much CO_2 was in Earth's water, rocks, and air?

Keeling realized that understanding these questions would require knowing the concentration of CO_2 in the air. Keeling started measuring CO_2 in Pasadena, California, near the university. His readings varied from day to day because of local CO_2 produced by cars and factories. Keeling moved his sampling equipment to Big Sur on the central California coast, far from urban CO_2 production. There he got much more uniform readings.

Keeling measured CO_2 concentration day and night. Soon he noticed that the CO_2 concentration was lowest in the late afternoon and highest in the early morning. Keeling figured out that plants were taking in CO_2 during the day and giving it off during the night. That explained the daily variation in CO_2 concentration.

Charles David Keeling

Mauna Loa Volcano

In 1956, Keeling moved his monitoring station to the top of Mauna Loa Volcano on Hawaii. He determined the volcano would be least affected by CO_2 from cars and plants. His first sample in March showed a CO_2 concentration of 313 parts per million (ppm). As he monitored CO_2 throughout the year, Keeling found that global CO_2 concentration varied on a seasonal cycle. The amount of CO_2 was highest in May (315 ppm) and lowest in October (311 ppm). The planet was "breathing," as Keeling put it. It was taking in CO_2 during the growing season in the Northern Hemisphere and giving off CO_2 in the fall and winter.

The most important thing Keeling discovered, however, was that the average concentration of global CO_2 increased every year. The Keeling data show a trend of increasing CO_2 concentration in Earth's atmosphere. The pattern in the CO_2 data is related to the burning of **fossil fuels**, like coal, gas, and petroleum. There is a direct relationship between CO_2 gas production by human activities and the concentration of CO_2 in Earth's atmosphere.

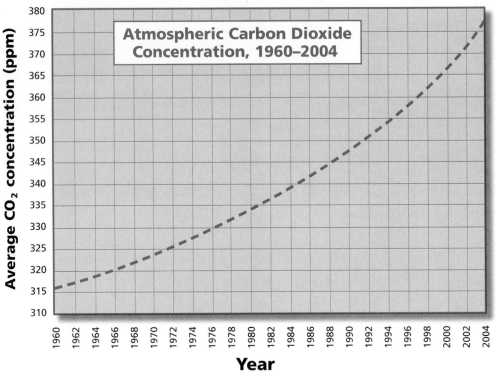

Keeling Curve

Atmospheric Carbon Dioxide Concentration, 1960–2004

The pattern in the data is known as the Keeling Curve. The curve gets steeper with each passing year. It is possible to use the curve to predict CO_2 concentrations into the future. Remember, when Keeling started measuring CO_2, the average global concentration was 313 ppm. Fifty years later the concentration was 378 ppm. If we continue to burn fossil fuels at the same rate, by the time we use them all up (in 100–200 years), the CO_2 concentration in the atmosphere will be around 1,500 ppm.

Carbon dioxide concentration in the atmosphere is important because CO_2 is a **greenhouse gas**. It absorbs energy and contributes to the warming of the atmosphere. This warming contributes to global **climate** change, which will have many effects on life on Earth. The Keeling Curve is a powerful and important scientific contribution to our understanding of the world and how it works.

Thinking about Carbon Dioxide

1. If humans started burning half as many fossil fuels as we did in 2004, what might the graph look like?

2. What might the graph look like if humans stopped burning fossil fuels all together?

The Frog Story

Scientists have noted a significant decline in the worldwide frog and toad populations in the last decade. The causes of the decline are difficult to pin down. There is evidence, however, that toxic chemicals may play a role.

Tyrone B. Hayes

Tyrone B. Hayes (1965–) is a biology professor at the University of California, Berkeley. He has studied frogs for most of his life. The first frogs he saw were in a swamp near his home in South Carolina. Now Hayes studies frogs across the planet, in Africa and North America.

Hayes found something strange going on with some of the frogs he studied. Frogs living in the wild were experiencing sex changes. The "male" frogs were producing eggs. The males were changing into females.

When Hayes and his team analyzed the frogs' environment, they found traces of a common **herbicide** (*herb* = plant; *cide* = kill) in the water. The concentration was only 0.1 parts per billion (ppb). That's one particle in 10 billion. But more tests in the lab showed that the herbicide was causing the changes.

The herbicide does not kill the frogs. But it does affect them. It changes their ability to reproduce. The substance in the frogs' environment, even in very low concentration, makes it impossible for the frogs to produce offspring. The pesticide kills the next generation of frogs, not the current one.

There are about 20,000 different pesticides used in the United States today. Each one is designed to kill unwanted organisms. But the pesticides don't all stay where they are applied. Particles of the substances are carried by wind and water to other locations. In other environments, the pesticides can have unintended results. This problem is growing all over the world, requiring government regulation and careful use of substances that might have an impact on natural environments.

The Bends

Hard-hat diving was invented in 1861. The diver climbed into a watertight suit with a brass helmet. An air hose was attached to the helmet. Air was pumped to the diver walking around on the bottom of the sea 20 meters (m) below the surface.

The **bends** is a condition that happened to deep-sea divers after returning to the surface. Divers felt dizzy, confused, and uncoordinated. They felt pain in their knees, hips, shoulders, and elbows. It became impossible for them to straighten their arms and legs. The pain caused divers to bend their arms and legs for pain relief. That's where the name bends came from.

The cause of the bends wasn't known until 1878. French scientist Paul Bert (1833–1886) figured it out. Nitrogen bubbles in the diver's blood and joints caused the bends. But where did the nitrogen bubbles come from? To answer that question, you need to know more about **saturated** solutions.

A solution is a solute dissolved in a solvent. We know about solids (salt) dissolved in liquids (water). Solutions can also be made when gases dissolve in liquids. That's what happens in the human body. Gases in the air that divers take into their lungs dissolve in the blood. Under normal conditions, the blood is saturated with dissolved nitrogen. No more nitrogen can dissolve.

When a diver goes underwater, the pressure increases. Pressure compresses the air in the diving suit. The air particles are pushed closer together. As a result, more air dissolves in the blood. Air is 78 percent nitrogen, so most of the additional gas that dissolves in the diver's blood is nitrogen. After the diver has been underwater for an hour, his blood is again saturated with nitrogen. But now it is saturated at high pressure, so there is more nitrogen in his blood than there was when he was at the surface.

Hard-hat diving helmet

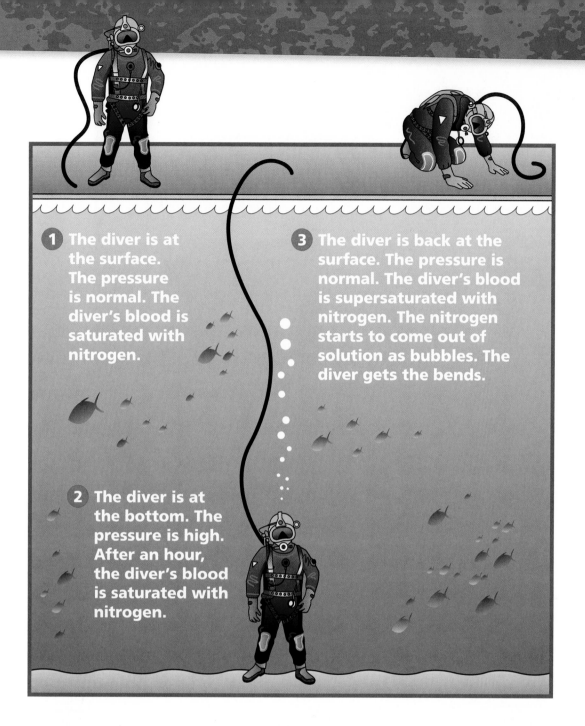

1 The diver is at the surface. The pressure is normal. The diver's blood is saturated with nitrogen.

2 The diver is at the bottom. The pressure is high. After an hour, the diver's blood is saturated with nitrogen.

3 The diver is back at the surface. The pressure is normal. The diver's blood is supersaturated with nitrogen. The nitrogen starts to come out of solution as bubbles. The diver gets the bends.

The trouble starts when the diver rises to the surface. The pressure drops to normal. The blood is holding much more nitrogen than it normally holds at surface pressure. The blood is **supersaturated** with nitrogen. The extra nitrogen comes out of solution (the diver's blood) as nitrogen bubbles. The bubbles get stuck in blood vessels and stop the flow of blood. Bubbles form in the fluids in joints, causing a lot of pain.

The bends is **decompression** sickness. Decompression means changing from higher pressure to lower pressure. That's when the diver feels the effects of too much nitrogen in the blood.

Caisson Disease

Decompression sickness also showed up in a different situation. In 1869, James Buchanan Eads (1820–1887) began building a railroad bridge across the Mississippi River. The bridge needed support in the middle of the river. This required a lot of digging underwater. How could that be done?

Eads used **caissons**. A caisson is a huge box with no bottom. It is placed on the bottom of a river with the open side down. Air is then pumped into the box. The air pushes the water out under the bottom of the box. Workers can dig and build foundations inside the caisson because it is filled with air.

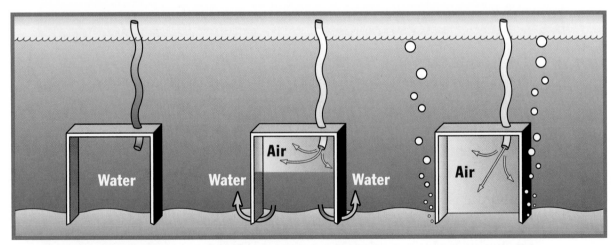

The caisson rests on the bottom of the river.

Pressurized air pushes water out under the bottom of the caisson.

Pressurized air keeps water out of the caisson.

The caisson is fitted with a tube that has tight-fitting doors. Workers can climb down the tube to the open Door 1 and go through it. They close Door 1 behind them. Then they can open Door 2 into the box. Using two doors maintains the pressure. This keeps the water from flowing back under the bottom of the box.

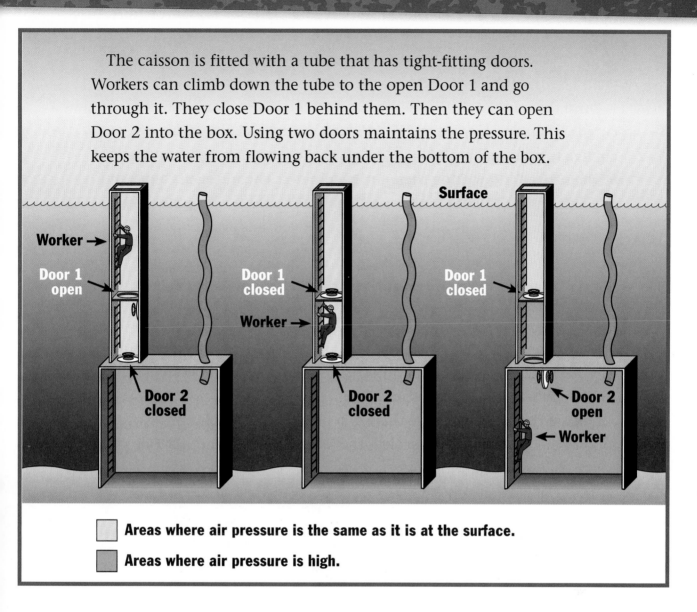

Surface

Worker →

Door 1 open

Door 2 closed

Door 1 closed

Worker →

Door 2 closed

Door 1 closed

Door 2 open

← Worker

☐ Areas where air pressure is the same as it is at the surface.

☐ Areas where air pressure is high.

The problem is the pressure. The pressure in the box has to be kept high enough to keep the water out. The workers are breathing concentrated nitrogen, so more nitrogen dissolves in their blood. At the end of a workday, their blood is saturated with nitrogen in the pressurized environment. When they return to standard atmospheric pressure at the water's surface, they have the same symptoms as deep-sea divers.

At a depth of 10 m underwater, the pressure is twice as high as standard atmospheric pressure. The workers are in no danger while they are working in the higher pressure in the caisson. The extra nitrogen in their blood does no harm. It is the change of pressure between the caisson and the surface that causes the extra dissolved nitrogen to rush out of the blood as bubbles.

Solving the Bends

Once the cause of the bends was understood, the condition was easily cured. Divers and workers in the caissons had to take more time to change the pressure back to normal. That meant coming halfway back to the surface and waiting there for 15 minutes. The nitrogen came out of solution slowly, so it didn't form bubbles. The extra nitrogen left the blood in the lungs and was exhaled. Then divers and workers could come to the surface safely.

Your Own Supersaturated Solution

Here's how you can observe the bends up close. Get a bottle of soda water. Soda water is a supersaturated solution of carbon dioxide dissolved in water. The high concentration of CO_2 stays in solution because the soda water is kept under pressure.

Look at the soda water. Are there any bubbles in the solution? No, as long as the soda water stays sealed, the CO_2 stays in solution. While you watch, twist the cap until you hear the "pfssst" that means the pressure has been released. Now what do you see? Bubbles!

As soon as the pressure is released, the extra dissolved CO_2 begins to come out of solution. It is the same with the bends. As soon as the diver comes back to the surface, the pressure drops, and nitrogen comes out of solution as bubbles.

A sealed bottle of soda water does not have any bubbles forming. Carbon dioxide bubbles rise from the bottle of soda water when pressure is released.

A Sweet Solution

Do you know what rock candy is? It is crystals of sugar. To make rock candy, you need to know about the science of solutions.

Sugar is a substance that dissolves in water. It takes about 100 grams (g) of sugar to saturate 50 milliliters (mL) of water at **room temperature**. Room temperature is 22° **Celsius** (C). But if you heat the solution, more sugar will dissolve. The hotter you get the solution, the more sugar dissolves in the water.

When the solution reaches its **boiling point**, it won't get any hotter. The boiling point of water is 100°C. When you see undissolved sugar in the pan of boiling sugar solution, you know the solution is saturated. There is about twice as much sugar dissolved in the boiling-hot saturated solution as there is in a room-temperature saturated solution.

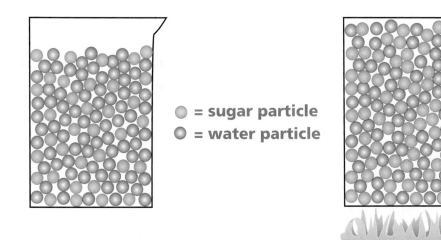

○ = sugar particle
◐ = water particle

A saturated solution of sugar at room temperature

A saturated solution of sugar at boiling temperature

What will happen to all that extra sugar when the boiling-hot saturated solution cools down? Will it stay in solution? Or will it come out of solution and pile up on the bottom of the container?

The sugar will stay in solution. A solution that has more solute than it should is a supersaturated solution. When the boiling-hot saturated sugar solution cools down, it is supersaturated.

Now the solution is ready to make rock candy. When you roll a wet string in sugar, the sugar sticks to the string. After the sugary string dries, it is covered with tiny sugar crystals.

When you put the string in the supersaturated solution, the extra sugar in the solution comes out of solution in the form of sugar crystals. The crystals stick to other sugar crystals stuck to the string.

The crystals will grow for a couple of days and then stop. Why do they stop growing? Sugar comes out of solution until the solution is no longer supersaturated. Then no more sugar comes out of solution.

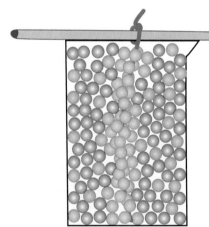

Sugar particles come out of solution on the crystals until the solution is again saturated at room temperature.

Sour Power

Do you like it when a glass of sour lemonade makes your lips pucker? If something that leaves a sour taste in your mouth is okay with you, then you're a fan of **citric acid**!

Citric acid is a white, solid acid. It is highly **soluble** in water. You can dissolve a lot of citric acid in a small amount of water. Like other acids, citric acid has a sour power. If you've ever bitten into a lemon, you know how citric acid tastes. About 6 percent of lemon juice is citric acid. Citric acid is found naturally in other fruits such as oranges, limes, and grapefruits. All of these fruits are known as citrus fruits.

Citric acid is a widely used food additive. It is sour, safe, and inexpensive. Every year tons of citric acid are added to foods for a number of purposes. It gives a sour tang to soft drinks and some candies. In jams and preserves, it helps fruits keep their color and flavor. Look on the backs of the cans and boxes in your cupboard. You will probably find a few packages that include citric acid as an ingredient. Citric acid is also used to make inks, dyes, medicines, and cleansers.

Scientists can extract citric acid from lemons and other citrus fruits. But because citric acid is so widely used today, huge quantities are needed. Factories have been built to produce citric acid by fermenting sugar.

Citric acid is not the only type of acid found in foods. Other weak acids occur in natural substances such as vinegar (acetic acid), tomatoes (acetic acid), and sour milk (lactic acid). There are also acids unrelated to food that are stronger and can be dangerous. They can burn through paper, cloth, and skin, so they must be handled with care. These strong acids are often used by scientists to conduct experiments and make chemicals.

East Bay Academy for Young Scientists

Have you ever wondered what it would be like to be a real scientist? Students in the East Bay Academy for Young Scientists (EBAYS) program don't wonder. They go out in the field and work as scientists. The program serves under-resourced communities in the San Francisco area in California. Some of the students started the program when they were in the sixth grade and stayed through middle school. Students in the program study soil, water, and air-quality problems in their communities. Their environmental work can help protect the families that live in their communities.

In the summer of 2008, students were enjoying the EBAYS summer program. That year it was held at Mills College in Oakland, California. During breaks on warm summer days, students would play in Leona Creek, which runs through the campus. A group of students decided to investigate the water quality of the creek for their summer research project. They first gathered samples of water from the creek. Water-quality kits were used to test the water samples in the classroom. Then they took these samples to the EBAYS laboratory at the Lawrence Hall of Science (LHS) to find how much lead each sample contained. LHS is a science and technology center at the University of California, Berkeley. Students found levels of lead above the legal limits of 15 parts per billion (ppb).

Water coming out of a sulfur mine

After discovering that the lead levels in the creek were so high, the students decided to test the water for other metals. They tested for iron and arsenic. They also wanted to find out how the metals were getting into the creek. They found the water's source at an abandoned sulfur mine. They collected some water samples at the mine and compared them to samples from the creek. They analyzed some of the water samples in the lab. They tested others on-site, using a test kit. Students found high levels of lead, iron, and arsenic at the mine. All the levels were above the legal limits. By comparing the water samples from the mine to those from the creek, the students saw that arsenic levels decreased as the water moved downstream. High levels of iron and lead were still found in these downstream samples.

Now that the students had some results, what was next? As part of the EBAYS project, they had to write a report and develop a poster. But this wasn't enough for the students. They wanted to do something more with the data. According to Elliot, one of the EBAYS students, "When we went to the mine, we were disgusted because it was this really contaminated area. That was when we started to think that we really needed to restore the creek. Water from the creek flows to a lake, and if the lake connected to the creek is polluted, then the water there can also be polluted. And since that water flows to the ocean, that spreads the pollution. Cleaning up one creek may not seem like much, but it is."

The students presented their results to the Oakland Department of Environmental Services. It turns out the mine owner had already been fined for not cleaning up the mine. The city was so impressed with the work the EBAYS students did that they were asked to adopt the creek. They were asked to begin work with other creeks in the area as well. Some of the EBAYS students volunteered for Creek to Bay Day and helped remove trash and nonnative plants.

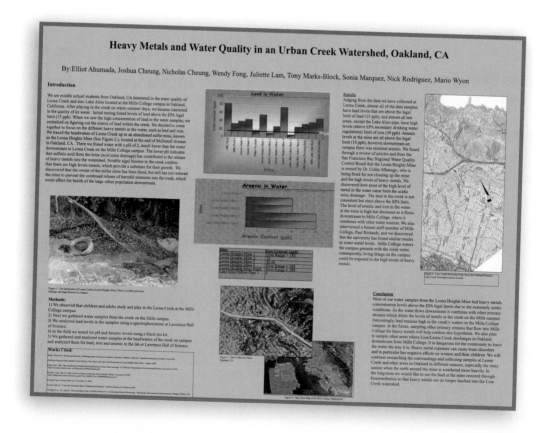

Elliot (center) and the other students presented their data and results at an American Geophysical Union (AGU) conference in San Francisco.

After the EBAYS summer program ended, the students wrote their findings. They created a poster that presented their results in a way that makes them easy for people to read. The students took their poster to the annual conference of the American Geophysical Union (AGU). At the conference, scientists could review the poster and discuss the findings. Elliot said, "At first, it was a little scary. We didn't see any other middle schoolers, but at the end, we started to get more and more confident. One guy congratulated us and said, 'You guys are really doing advanced stuff.' After that, we started presenting more, and we got a lot more confident." It turns out they were some of the youngest students ever to present at an AGU conference!

The AGU conference was just the beginning for these students. They submitted a research paper related to their work in Oakland to a contest encouraging sustainable development. Their paper won the contest. On May 14, 2010, the students presented their work to the Commission on Sustainable Development at the United Nations headquarters in New York.

Drinking Ocean Water

The world's demand for drinking water is increasing. The human population is over 7 billion. Climate patterns are changing and causing droughts. As a result, people are looking for new sources of drinking water. We already use surface water such as lakes, rivers, and streams. And we pump water from the ground. Where else can we look for drinking water?

The ocean holds 97 percent of Earth's water. Can we use the ocean for drinking water? What do you think?

Ocean water contains dissolved salts. It has a high concentration of sodium chloride. Getting drinking water from ocean water requires separating the water from the salt. This process is called **desalination**.

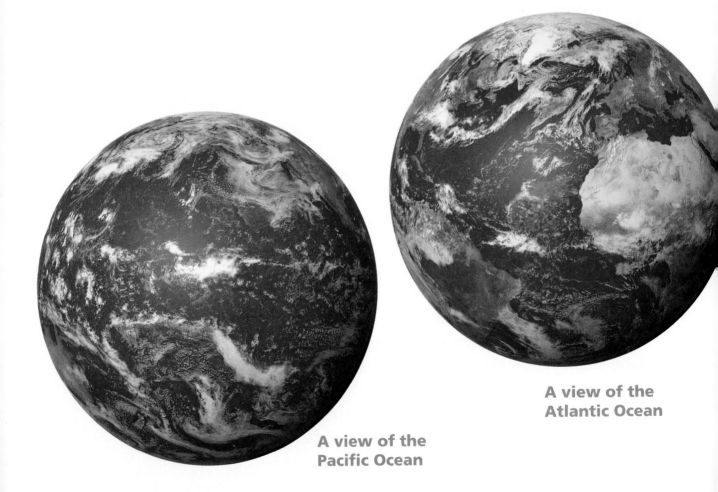

A view of the
Atlantic Ocean

A view of the
Pacific Ocean

Desalination

How did you separate water and salt? Could that method make drinking water? You evaporated the water, leaving the salt behind. The liquid water became water vapor. It escaped into the air. But now we want to hold onto the water.

An evaporation system with a condenser is possible. A condenser changes water vapor into liquid water. But heating enough water for a community would be very expensive. Evaporation may not be a good design for desalination.

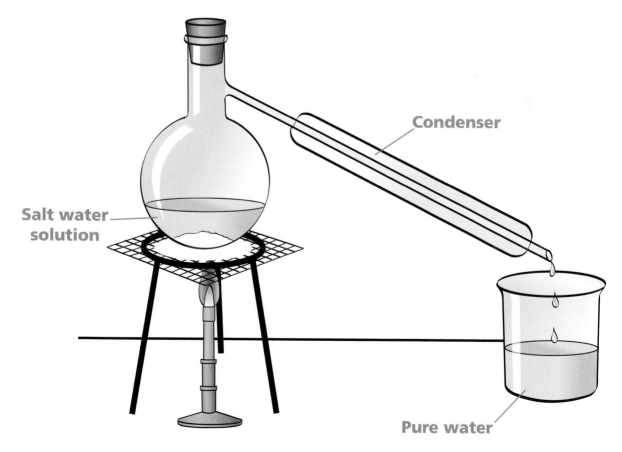

Condenser

Salt water solution

Pure water

Engineers have developed another solution for desalination. The process uses a filter called a **semipermeable membrane**. If something is **permeable**, Substances can pass through a permeable membrane. For instance, a T-shirt is a permeable membrane. If you stretch a T-shirt across the top of a bucket, you can pour water into the bucket through the T-shirt. Plastic bags are **impermeable**. If you stretch a plastic bag across the top of a bucket, water cannot enter the bucket. A semipermeable membrane allows water to pass through. It blocks dissolved materials.

Look at this setup. A semipermeable membrane is in the bottom of a glass U-tube. A blue salt solution is in one side of the tube. An equal volume of pure water is in the other side. Observe what happens after 30 minutes.

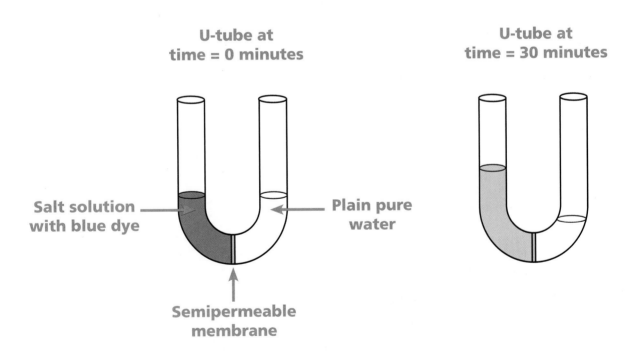

U-tube at time = 0 minutes

U-tube at time = 30 minutes

Salt solution with blue dye

Plain pure water

Semipermeable membrane

What happened to the two solutions after 30 minutes?

The volume of salt solution increased, and the volume of pure water decreased! The membrane is permeable to water molecules. It is impermeable to salt and dye molecules. Water molecules passed through the membrane into the salt solution. The salt solution became less concentrated. This movement of water molecules is **osmosis**. A force drives the water molecules through a semipermeable membrane. This force is **osmotic pressure**.

Here, a pump applies pressure to the salt solution in the U-tube. The pump pressure acts against osmotic pressure. The pump pressure stops water molecules from passing into the salt solution. Say you increased the pump pressure even more. The pressure pushes water molecules out of the salt solution. They move into the pure-water side of the tube. Removing water makes the salt solution more concentrated.

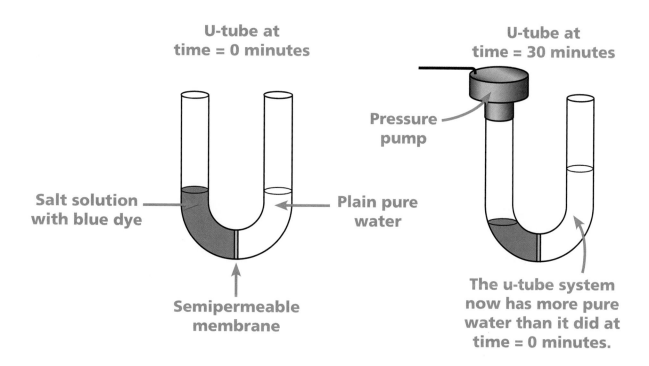

U-tube at time = 0 minutes

U-tube at time = 30 minutes

Salt solution with blue dye

Plain pure water

Pressure pump

Semipermeable membrane

The u-tube system now has more pure water than it did at time = 0 minutes.

Reverse osmosis filters at a water utility plant

The permeability of the membrane does not change under pressure. It still allows only water molecules to pass through. Using high pressure and a semipermeable membrane pushes water out of a salt solution. This process is **reverse osmosis**.

A semipermeable membrane is a special filter. Filters let some things pass through and hold other things. A screen can separate gravel from smaller materials. And a paper filter can separate powder from a salt solution. But a paper filter cannot separate salt from water. You had to use evaporation to separate it.

Criteria and Constraints

Millions of people live near the salty ocean. Reverse osmosis is one way to get water to a thirsty world. Does it meet the **criteria** and **constraints** for a successful solution?

The criteria include producing lots of pure drinking water. The usual constraints are cost and availability of materials. Semipermeable membranes are delicate and expensive. And reverse osmosis requires energy to make pressure. Reducing the energy costs is critical. Semipermeable membranes must be tough and perform well at low pressure.

Another constraint is disposal of waste products. Desalination produces salt as waste. How will we dispose of it?

A water desalination plant in Dubai

Conservation and Recycling

Desalination is one way to get more drinking water. Other ways include conservation and recycling.

Conservation involves sensible practices and engineering solutions. Landscaping with drought-tolerant plants is a sensible practice. We can replace green lawns with native plants that don't need much water. We can use water-efficient appliances. Engineers have designed low-flow washing machines, dishwashers, showerheads, and toilets. Engineers have also designed better ways to irrigate crops.

A lot of water used in our homes and businesses becomes waste water. It ends up in the sewer system. Engineers are developing systems for recycling this waste water. Modern sewage treatment returns clean water to natural environments. Secondary water systems can produce water that is clean enough for watering plants and washing cars. Reverse osmosis can treat waste water and make it safe to drink.

Thinking about Drinking Ocean Water

Look at the environment in each photograph. Why might desalination work well in these environments? Can you think of other environments that might be good for desalination?

Creative Solutions

People face problems every day. Problems may be small or big, from doing homework to curing cancer. Some problems make people work harder than they would like to work. But while some people complain, others act. Inventors are people who solve problems in creative ways. They don't turn away from problems. Inventors look for solutions. They might be scientists in labs. They might be ordinary people at home. Even kids are inventors! Anyone can be an inventor.

The Flying Monkeys

A team of Girl Scouts in Ames, Iowa, called the Flying Monkeys invented a device to help a three-year-old girl named Danielle, who was born without fingers on one hand, write for the very first time.

The Flying Monkeys

Their invention, called the BOB-1, is a prosthetic hand device that helps people hold and grip objects, such as a pencil. The device is intended to help people born with limb differences lead normal and healthy lives.

The Flying Monkeys won the FIRST LEGO League Global Innovation Award and $20,000 in 2011. They also have a short-term patent on their invention.

Arthur Fry

Sometimes inventors have goals in mind. They want to make life easier. They consider a problem, and then they work until it is solved. Arthur Fry (1931–) is such an inventor. Fry sang in his church's choir. He marked pages in his hymnal with scraps of paper. But they fell out as Fry turned from one hymn to another.

Luckily, Fry knew of another inventor's experiment. Spencer Silver had been experimenting with adhesives. He made one that stuck things together and allowed them to be pulled apart without damage. The adhesive could even be reused. No one could think of a use for Silver's invention except Arthur Fry.

Arthur Fry with his invention

Fry tried some of Silver's weak adhesive on his paper scraps. It wasn't perfect. The book pages were sticky after the scrap was removed. Fry experimented for 18 months. He worked until the adhesive was just right. Then he made a machine to apply the adhesive to paper. He called his invention Press and Peel.

Fry showed Press and Peel to his bosses at the 3M Company. They weren't impressed. They couldn't imagine anyone needing sticky note paper. But Fry convinced them to let 3M secretaries try his invention. Soon free samples were sent to other offices. People were enthusiastic! They stuck notes in books and on walls. They thought of many uses for Fry's invention. 3M renamed the product Post-it® notes. By 1980, Post-it notes could be bought all over the United States. Today, Post-its are a best-selling office supply, all because Fry's paper scraps fell out of his hymnal. Arthur Fry worked until he had solved his problem.

Goodyear accidentally discovered how to make rubber strong enough for tires and many other useful things.

Tire swings are one way to recycle Goodyear's invention.

Charles Goodyear

Although most inventions require a lot of hard work, some inventions happen by accident. Charles Goodyear (1800–1860) was interested in rubber. Rubber is made from the sap of rubber trees. It bounces and can erase pencil mistakes. But natural rubber melts in hot weather, and it shatters in cold. Goodyear thought he could make rubber better. He tried mixing it with other materials. None of them made rubber more useful.

Then in 1839, Goodyear dropped rubber and a chemical called sulfur onto his stove by accident. Cleaning up the mess, he found heat had changed the rubber. It was suddenly firm. High temperature didn't melt it, and it didn't shatter when it got cold. Goodyear had made rubber better by accident! He called the heating process vulcanization. Because of Goodyear's accident, now rubber is used to make tires, shoe soles, balls, and many other useful things.

An inventor once said, "When I see something that I don't like, I try to invent a way around it." Inventors not only see what is wrong, but they also see how things can be made better.

George de Mestral

Swiss engineer George de Mestral (1907–1990) loved to hunt. But often he and his dog returned from the woods covered with burrs. They stuck to de Mestral's clothes and to his dog's fur. It took a long time to pull them out. De Mestral didn't understand why the burrs were so hard to remove. He put a burr under his microscope to take a closer look. He saw that it was covered with tiny hooks. They were perfect for grabbing onto fabric and hair. De Mestral wondered if he could use such hooks in a helpful way. Could they hold things together?

De Mestral set out to imitate the burr. He imagined a hook-and-loop system. It would be able to do the work of zippers, snaps, laces, and buttons. The main difficulty was finding the right material. De Mestral wanted to use nylon, a strong material. But nylon was so tough that no one could cut it into hooks. In the end, de Mestral had to invent a special cutting machine for the nylon just so he could use it to produce his invention.

De Mestral called the hooks and loops Velcro® after two French words, *velour* (velvet) and *crochet* (hook). Production of Velcro began in the mid-1950s. Today, Velcro is used on things such as shoes, sports equipment, clothing, toys, backpacks, and watchbands. Velcro has even been used by astronauts to anchor objects in the space shuttle while in space!

De Mestral's hook-and-loop system, which he called Velcro

Velcro is used on shoes and many other things.

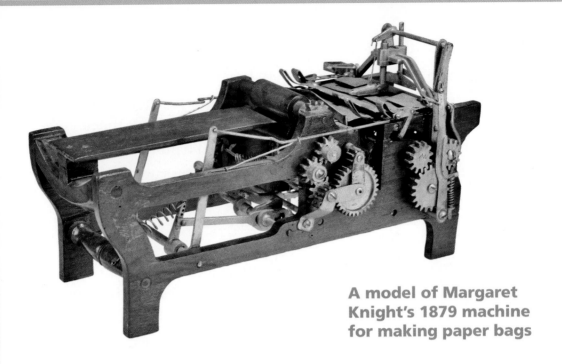

A model of Margaret
Knight's 1879 machine
for making paper bags

Margaret Knight

Margaret Knight (1838–1914) worked in a cotton mill when she
was a child. One day, she witnessed an accident. Cotton strands
running through an automated loom became tangled, causing a
steeltipped shuttle to fall out of the loom and hit a worker. Knight
went home upset but thoughtful. Only 12 years old, she
designed a stop-motion device. It made the loom shut
down when strands became tangled. Her invention
helped keep workers safe.

Knight worked at many jobs as a young
woman. At a paper bag company, she became
interested in making bags that were more
efficient to pack, as well as automating
the way they were made. She spent time
observing the bags being made by hand.
Then she invented a machine to cut, fold,
and glue flat-bottomed bags together.

Knight wanted to get a patent for her machine. A patent is a document from the government that credits the inventor for an invention. But when Knight tried to get her patent, she discovered that someone else had stolen her idea! Charles F. Annan had seen Knight's model at a machinist's shop and copied it. Knight fought him in court. She showed her plans, as well as diary entries where she wrote about her work. Knight won the patent. Today's paper bags are still made using her ideas.

Knight never had a formal education. She credited her achievements to a love of tools and machines. In her lifetime, she received 27 patents. She improved automobile motors and machines that made textiles, shoes, and tin cans. Knight was inducted into the National Inventors Hall of Fame in 2006, 92 years after her death.

National Inventors Hall of Fame

The National Inventors Hall of Fame was started in 1973 and is currently located in Alexandria, VA. It's mission is to "honor the women and men responsible for the great technological advances that make human, social, and economic progress possible." As of 2015, there are 516 inventors honored in the Hall of Fame. Each year in February, inventors selected by a panel of experts are inducted into the Hall. Only inventors that have had a US patent that has promoted the progress of science and technology and improved the welfare of society are considered.

Stephanie Kwolek

Stephanie Kwolek (1923–2014) is a **chemist** who worked in a laboratory to create new materials that are useful to society. She started working for the Textile Lab at DuPont® soon after graduating from college in 1946. She thought this job would be temporary so that she could earn enough money to attend medical school. But she really liked the work. She spent her entire career working as a research chemist.

Stephanie Kwolek

Kwolek's job was to develop new synthetic polymers. Polymers are long molecules that can be made into fabric or plastics. In the 1960s, she worked with a team of other chemists to develop high-performance fibers that were strong and stiff. One of the goals of this research was to find a replacement for the steel in steel-belted tires. If the steel could be replaced with a lighter-weight but strong material, then the vehicles would use less gasoline to run. The team also researched fibers that wouldn't melt at very high temperatures.

In her lab, Kwolek mixed substances to make a polymer. Then she melted the polymer into a liquid. The next step in the process was to spin the liquid in a machine called a spinneret. The spinneret turned the liquid into fibers. Other scientists on the team tested the newly created fibers to find their mass, how strong they were, and to determine other properties such as how easily the fibers would stretch or break.

One day, Kwolek created a polymer that would not melt, even at very high temperatures. She decided to see if she could get the polymer into a liquid state by dissolving it in a solvent, rather than melting it with heat. She finally found a solvent that would dissolve the polymer, but the resulting solution was different from any other polymer solution she had observed. Instead of being viscous and clear, like syrup, this new solution was more like water but cloudy. No one expected fibers to form by spinning the liquid in the spinneret. But Kwolek insisted on giving it a try. The result surprised everyone in the lab. The newly created synthetic fibers were much stiffer and stronger than any created before. Kwolek had discovered a new fiber called an aramid fiber. She also discovered a new type of substance called liquid crystalline solution.

The new fiber was named Kevlar®. It is five times stronger than steel (gram for gram), and about half the density of fiberglass. It took almost 10 years from the development of the fiber to the marketing of the first product. Can you name some products made from Kevlar?

One of the most popular uses for Kevlar is bullet-proof vests. These vests, worn by police and other law enforcement officers, save many lives. Kevlar is used to make skis, safety helmets, and bicycle tires. It is used in automobile tires and brake pads, and in cables used for suspension bridges. Kevlar is also used as a protective outer cover for fiberoptic cables. The shells of spacecraft, airplanes, and boats are made out of Kevlar. You can even find Kevlar in some audio speakers and on some smartphones.

Ask a Chemist

Beryl Baker

Beryl Baker's teacher asked her students to do a career report for an assignment. Beryl decided to do her report on a chemist's career. To find out what chemists do, she talked to Angelica Stacy, professor of chemistry at the University of California, Berkeley. Beryl had this interview with Professor Stacy.

BB We are studying mixtures and solutions in our class. My teacher said people who work with mixtures and solutions are called chemists. What does it mean to be a chemist, and what does a chemist do?

AS Those are good questions. Chemists do study mixtures and solutions, but they also study all other states of matter, including gases and solids. We study their properties and try to find out why they do what they do. We then use that information to make new things.

Chemists make many of the things around you: the material of your jacket, the dye in your jeans, medicines, plastics, and lots more.

BB Why do you like chemistry?

AS It's more that I like science, and chemistry is one piece of science. For me, chemistry is a way of thinking about the world and contributing to society at the same time.

Angelica Stacy

BB Was there a person who started you thinking about chemistry?

AS My father. He was an engineer with RCA. He never got a college degree because his family was poor, so he taught himself most of what he needed to know. Because of him, there was always science in our house. He was always making things that interested me and talking about how sound and electricity worked.

BB How long did you go to college to learn about chemistry?

AS Four years of college as an undergraduate to get my bachelor's degree, and then 4 years of graduate work to get the advanced degree so I could become a professor. In my case, it took 8 years of college-level study.

BB Do all chemists work at colleges or universities?

AS No. There is a big chemical industry worldwide. One [industry] that you probably know about is the medicine companies that create and manufacture medicines to fight disease, relieve pain, treat injuries, and so on. Lots of chemists work in this industry. Chemists work in the food industry, enriching and preserving the things we eat. Agriculture depends on chemistry. All those things in the bathroom were developed by chemists: soaps and detergents, cleansers, stain removers, shampoo, hair color, mouthwash, deodorant, toothpaste. Chemists are at work in many industries and government agencies.

BB Does everybody who studies chemistry become a chemist?

AS People may use their understanding of chemistry to launch into other interesting professions. One chemist I know is now a multimedia designer. His chemistry training prepared his mind for the demanding work of creating computer programs that help students learn about science subjects in interesting and fun ways. Others might move into university administration, law enforcement, government work, medicine, or business. And many students who start in chemistry move on to other fields of science and engineering because chemistry is recognized as the basic science. Knowledge of matter and its fundamental behaviors is a good place to start with any career in science.

BB What tools and instruments do you use when you are doing your chemistry work?

AS We use lots of different instruments to help us find out what things are made out of and how they are put together. As it turns out, some of our most important tools use light to give us information. The way different wavelengths of light are absorbed, reflected, or changed by substances tells us a lot about the structure of those substances.

Separation techniques are important, too. You undoubtedly used evaporation as a way to separate a solvent from a solute in your study of mixtures and solutions. You may have used chromatography to separate the pigments in inks. We do the same kinds of things when we are confronted with complex mixtures of substances. This part of chemistry is called analysis: finding and identifying all the parts in mixtures.

BB Have you invented or discovered anything?

AS My specialty is materials. I was investigating superconductors. Superconductors conduct electricity without resistance. But now I have turned my attention to trying to develop a new kind of refrigerator. Most refrigerators today use a gas called freon. But there is a problem with freon. It damages the ozone layer in the atmosphere, so freon is being phased out. I'm trying to develop a solid refrigerant. When you run electricity through it, it gets hot on one end and cold on the other. Right now it is only 10 percent efficient. Your freon refrigerator is 30 percent efficient. The material must get three times better to be useful on a broad scale. This is what I am trying to discover in my lab. But breakthroughs don't come easily. Discovery is hard work.

BB What are the most interesting things you get to do?

AS Teaching [is one]. My refrigeration research with graduate students is a kind of high-level teaching. I also teach basic chemistry for undergraduates, and I'm developing new ways to teach chemistry at the high school level. Another interesting part of my work is sharing ideas with other chemists all over the world. I have friends and colleagues in many countries as a result of my work in chemistry.

BB Is there anything else you would like to say to students about chemistry?

AS There are lots of opportunities in chemistry and in science generally. Whatever your interests are, knowledge of science can be part of your plans. I'd like to remind girls particularly not to be intimidated by science. In science, you have the opportunity to find things out that might help to solve problems, like my work in refrigeration, or disease control, food production, [or] lots of other things. It's a good feeling to contribute to the knowledge of the world.

When Substances Change

Tina made two solutions. She mixed citric acid and water in one cup. She mixed baking soda and water in another cup.

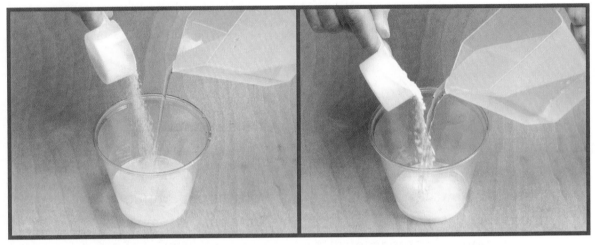

Citric acid and water   **Baking soda and water**

Tina poured the citric acid solution and the baking soda solution into an empty cup. At the same time, her brother Leo opened a bottle of soda water and poured some into an empty cup.

Combining citric acid and baking soda solutions   **Pouring soda water**

The liquids in both cups bubbled and fizzed. In 1 minute, they settled down to a slow, steady stream of bubbles. After 15 minutes, both cups were clear and still.

What happened in the two cups? Tina thought there was a chemical reaction in both cups. When you mix two substances and a change occurs, the change is evidence of a chemical reaction. Tina saw bubbles in both cups. Bubbling is a change. Tina thought there must be reactions going on in both cups.

Leo had a different idea. He knew that the bubbles in Tina's cup were filled with carbon dioxide (CO_2) gas. He observed that there were no CO_2 bubbles in the citric acid solution or the baking soda solution before they were mixed. The CO_2 bubbles formed only after the two solutions were mixed. Carbon dioxide gas was a new substance, evidence of a chemical reaction.

But Leo wasn't sure about the soda water. He didn't mix the soda water with any other substance. He just opened the bottle and poured. And up came the CO_2 bubbles. He didn't think the bubbles in the soda water were the result of a chemical reaction. But where did the CO_2 come from?

Remember, CO_2 dissolves in water. At the soda water bottling plant, water is saturated with CO_2 under high pressure. Leo released the pressure by removing the bottle cap. At the moment the pressure dropped, the solution became supersaturated.

The extra CO_2 then came out of solution in the form of bubbles. Carbon dioxide was not a new substance, so there was no chemical reaction.

Reactions That Don't Fizz

In 1772, Swedish pharmacist
Karl Wilhelm Scheele (1742–1786) heated a
sample of red-brown mercury oxide powder
in a test tube over a flame. He observed a
gas coming up and a liquid metal at the
bottom of the test tube. The mercury oxide
powder was gone. In its place were two new
substances, a liquid metal and a gas.

Scheele observed a chemical reaction.
A single **reactant**, mercury oxide, had
changed into two new **products**, mercury
metal and oxygen gas. By conducting this
reaction, Scheele isolated and described the
element oxygen.

Think about a rusty nail on the sidewalk,
a gas burner on a stove, and a magnificent
Fourth of July fireworks display. What do
they have in common? They are all reactions
in progress.

The Rusty Nail

A nail left outdoors in the environment gets rusty. Why is that?
What is rust?

The nail contains iron. The iron in the nail reacts with oxygen
in the air. The reaction produces a new product that we call rust.
The scientific name for rust is iron oxide.

The reaction is very slow. It takes days
or weeks for rust to become visible
on the nail. It might take a hundred
years for the whole nail to
change into iron oxide. At
the end of that time, the iron
nail is gone. In its place is the
new substance iron oxide.

The Gas Burner

Natural gas occurs in large quantities underground. It is collected and delivered through pipes to homes and businesses. When the gas is delivered to a stove or furnace, it is burned to produce heat. The heat cooks food and warms the air.

Gas changes into heat. How does that work? Natural gas is made of carbon and hydrogen. When the gas (one reactant) is mixed with oxygen from the air (the second reactant), a reaction is ready to happen. But nothing happens until the reaction is started with a spark or flame. Once the reaction starts, it will continue by itself until one of the reactants is used up. If the natural gas is used up, the reaction stops. If the oxygen is used up, the reaction stops.

While the natural gas flame is burning, the reaction is happening. Three things are given off by the reaction, two products and heat. The products are CO_2 and water. Here's one way to illustrate the reaction.

The major ingredient in natural gas is a substance called **methane**. The methane particle is made of carbon and hydrogen and looks like this.

The oxygen particle in the air looks like this.

In the burning flame, each particle of methane reacts with two particles of oxygen. The methane particles and the oxygen particles break apart and form new particles of CO_2 and water. This is how the reactants change to form the new products.

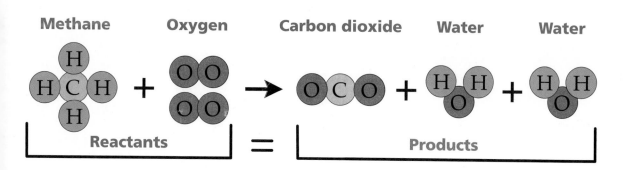

77

You can see that matter is conserved during the reaction. The number of carbon particles, hydrogen particles, and oxygen particles is the same on both sides of the arrow. In other words, the matter in the reactants is exactly the same as the matter in the products.

The natural gas reaction is a fast reaction. The change from reactants to products occurs in a flash. The products are both gases, CO_2 and water vapor, so the reaction is "clean." The only concern is the waste product CO_2, which enters the air.

Fireworks

Another fast reaction is a stunning fireworks display. This kind of reaction is called an **explosion**. To qualify as an explosion, the reaction must happen very fast and must produce light, heat, and sound energy, plus a lot of gas. Because the gases expand so rapidly, explosions come with a loud kaboom. The people who design fireworks know what substances to put into each charge to produce different colors. The green color is the product of one substance, the red color is from another substance, and so on. The result is a thrilling experience for your eyes and ears.

Air Bags

The automotive air bag was invented in 1952 as a safety device for people. Twenty years later, air bags started to appear in American cars as an extra. Today, all cars sold in the United States have air bags in front, one for the driver and one for the passenger. Many cars have additional air bags in the ceiling and doors.

An air bag is a fabric bag that inflates like a big balloon the moment a car crashes into something. The bag has to inflate fully in a few thousandths of a second! How is that possible?

It's a chemical reaction. When a car smacks into a solid object, sensors in a triggering device start the action. A pulse of electricity flows to the igniter, and a wire gets hot. The hot wire starts a very fast reaction, which produces a large volume of gas, usually nitrogen. The expanding gas bursts open the steering wheel or dashboard, and the bag pops out. It has to be fully inflated before the driver's or passenger's head and chest reach the steering wheel or dashboard. That's fast inflation!

The air bag is full of carefully designed holes. When the person's body contacts the bag, the force of the impact squeezes air out of the holes. The air bag doesn't stop the forward motion of the person. It slows the speed of the forward motion. This is important because the longer it takes for the person's body to come to a stop, the less he or she will be hurt by the collision.

The inflation reaction is pretty close to being an explosion. The definition of an explosion is a fast reaction that produces gas, heat, light, and sound energy. The air bag reaction is fast, produces a lot of gas, gets warm, and makes a bang, but has no flash of light. It is close to an explosion, but not quite. Still, the action is forceful. People have been injured by air bags. The benefits, however, far outweigh the hazards. Thousands of lives have been saved by air bags since they were first installed in cars more than 40 years ago.

An air bag inflates completely in 20–30 milliseconds.

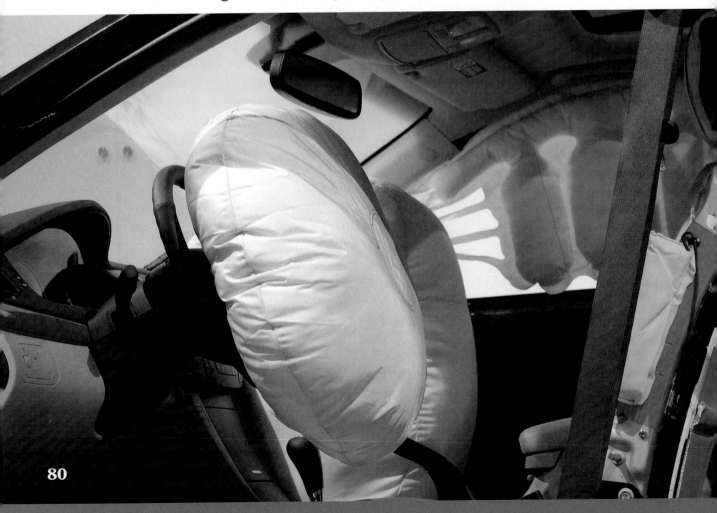

Celsius and Fahrenheit

Celsius

°C

Celsius and Fahrenheit are two **scales** used to measure temperature. Both scales are based on the **freezing point** and boiling point of pure water at sea level. The Celsius scale has 100° between the two points. The Fahrenheit scale has 180° between the freezing point and boiling point.

Today most countries use the Celsius scale to measure temperatures. The United States, however, still uses the Fahrenheit scale.

Anders Celsius

The Celsius scale is named for Anders Celsius, a Swedish astronomer. Celsius lived from 1701 to 1744. In 1742, he created a temperature scale. This scale used 0°C to mark water's boiling point and 100°C to mark its freezing point. A few years later, another scientist changed Celsius's scale so that 0°C was the freezing temperature and 100°C was the boiling temperature. Celsius's scale was originally called the centigrade scale. It was renamed in the 1940s to honor the inventor.

Daniel G. Fahrenheit

The Fahrenheit scale is named for German scientist Daniel G. Fahrenheit. Fahrenheit lived from 1686 to 1736. In 1714, he invented the first mercury thermometer. He invented a temperature scale to go along with it. Fahrenheit's thermometer marked normal human body temperature as 98.6°F.

Fahrenheit thought he had found the lowest possible temperature by mixing ice and salt. He set the temperature of this mixture at 0°F. Then he set the freezing point of water at 32°F. He also set the boiling point of water at 212°F.

Body temp 37°

Room temp 22°

Freezing point of water 0°

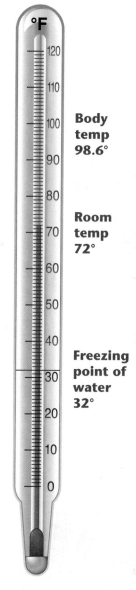

Fahrenheit

°F

Body temp 98.6°

Room temp 72°

Freezing point of water 32°

Science Safety Rules

1. Listen carefully to your teacher's instructions. Follow all directions. Ask questions if you don't know what to do.

2. Tell your teacher if you have any allergies.

3. Never put any materials in your mouth. Do not taste anything unless your teacher tells you to do so.

4. Never smell any unknown material. If your teacher tells you to smell something, wave your hand over the material to bring the smell toward your nose.

5. Do not touch your face, mouth, ears, eyes, or nose while working with chemicals, plants, or animals.

6. Always protect your eyes. Wear safety goggles when necessary. Tell your teacher if you wear contact lenses.

7. Always wash your hands with soap and warm water after handling chemicals, plants, or animals.

8. Never mix any chemicals unless your teacher tells you to do so.

9. Report all spills, accidents, and injuries to your teacher.

10. Treat animals with respect, caution, and consideration.

11. Clean up your work space after each investigation.

12. Act responsibly during all science activities.

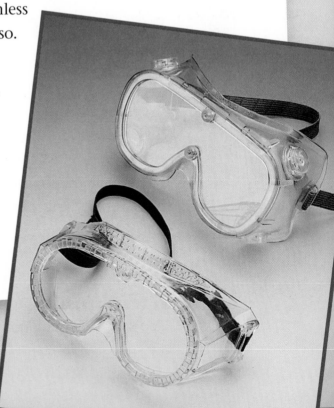

Glossary

atmosphere the layer of gases surrounding Earth (air)

bends a condition that causes pain in deep-sea divers' arms and legs after they return to the surface

boiling point the temperature at which liquid turns to gas. The boiling point of water is 100°C.

caisson a large box with no bottom used as a working environment underwater.

carbon dioxide (CO$_2$) a compound made from carbon and oxygen

change to make different

chemical reaction the process in which substances change to make one or more new substances that have different properties from the original ones

chemist a person trained in chemistry

citric acid a white, crystalline solid acid found in citrus fruits

climate the average or typical weather conditions in a region of the world

compress to reduce in volume by applying pressure

concentrated a solution containing a lot of solute

concentration the ratio of solute to solvent in a solution

condensation the change of state from gas to liquid

conserve to stay constant during an interaction. Matter can change, but it's always conserved.

constraint a restriction or limitation

criteria a set of standards for evaluation or testing something

crystal the natural form of some solid substances. Crystal shape is also a physical property that helps to identify a substance.

decompression the change from higher pressure to lower pressure

Degree Celsius (°C) the unit of temperature in the metric system. Water freezes at 0°C and boils at 100°C

density mass per unit volume

desalination the process of separating the water from the salt in ocean water

diatomaceous earth the shell-like remains of microscopic aquatic organisms (diatoms)

dilute a solution containing little solute

dissolve the process of a substance becoming incorporated uniformly into another

energy the ability to make things happen. Energy can take many forms, such as heat and light.

engineer a scientist who designs and builds systems that result in new products or solve problems

evaporate when water in a material dries up or goes into the air or the change of state from liquid to gas

evaporation the change of state from a liquid to a gas

explosion a fast reaction that produces heat, light, and sound energies, and a lot of gas

extract a solution of substances dissolved out of organic material

fossil fuels the preserved remains of organisms that lived long ago and changed into oil, coal, and natural gas

freeze the change of state from liquid to solid as a result of cooling

freezing point the temperature at which a liquid becomes a solid. The freezing point of water is 0°C.

gas a state of matter that is shapeless and expands to fill any closed container it is placed in

gaseous existing in the gas state (not a solid or liquid)

greenhouse gas a gas, such as carbon dioxide, that contributes to the warming of the atmosphere

herbicide a poison intended to kill plants

impermeable not allowing substances to pass through

liquid a state of matter that flows and takes the shape of the container it is in

magnetism the ability to attract iron

mass a quantity of matter

matter anything that has mass and takes up space

melt to change state from a solid to a liquid state as a result of warming

methane the main ingredient in natural gas

mixture two or more materials together

model an explanation or representation of an object, system, or process that cannot be easily studied

nitrogen a colorless, odorless gas that makes up about 78 percent of Earth's atmosphere (air)

oceanographer a scientist who studies the chemistry and geology of the ocean

osmosis the movement of water through a semipermeable membrane in a solution with a higher solute concentration until the concentration on both sides of the membrane is equal

osmotic pressure a force that drives water molecules through a semipermeable membrane

oxygen a colorless, odorless gas that makes up about 21 percent of Earth's atmosphere (air)

particle a very small piece or part

permeable allowing substances to pass through

physical property a characteristic that describes a substance, such as color, size, shape, or texture

pressure the force exerted by an object or fluid on something else

product the substance(s) created or resulting from a chemical reaction

radioactivity the process by which certain elements emit radiation

rain liquid water that is condensed from water vapor in the atmosphere and falls to Earth in drops

ratio the comparison of two (or more) parts of a whole

reactant one of the starting substances in a chemical reaction

reverse osmosis the process of using pressure to push pure water out of a salt solution using a semipermeable membrane

room temperature the temperature of a comfortable indoor space. Room temperature is about 22°C.

salt table salt. A white crystalline solid that dissolves in water. Its chemical name is sodium chloride.

saturated a solution in which the solvent cannot dissolve any more solute

scale something divided into regular parts to use as a tool for measuring. Rulers and thermometers both use scales.

semipermeable membrane a filter that allows water to pass through it but does not let dissolved materials pass through it

sodium chloride the chemical name for table salt

solid a state of matter that has a definite shape

soluble capable of being dissolved. Table salt is soluble in water.

solute a substance that dissolves in a solvent to form a solution

solution a mixture of one or more substances dissolved in solvent

solvent a substance in which a solute dissolves to form a solution

substance a pure material that is not a mixture

supersaturated a solution that contains more solute than it normally would at a given temperature or pressure

temperature a measure of how hot or cold matter is

transfer to pass from one place to another. Heat energy transfers to make solids melt.

transparent clear; describes something through which you can see an image clearly

volume three-dimensional space

water vapor water in its gas state

Index